Table of Contents

The Spirit Diggers

Chapter 1
Mind-set

Until recently, I had always said, "If I can't see it, it isn't real!" Being a stubborn male with a scientific education didn't help matters either. Although I always romanticized ghost stories and the thought of being scared out of my mind was very exciting for me, I still had trouble believing that ghosts did exist. That is, until an overnight investigation at a Western New York asylum which was built in the 1700's. This location is well known for it's paranormal activity and even though I didn't have too much happen during the actual investigation, most of the other team members did and the evidence we collected was

Denise Harris
Lead Investigator / E.V.P. Specialist

Mind-set

indisputable. I believe that your mindset entering an investigation will determine if you have a positive experience when it comes to witnessing paranormal activity. I have witnessed other team members have personal experiences while I, being 2 feet away, have absolutely nothing happen to me. Being very frustrated, I couldn't help but think that it had to be imaginations being slightly overactive. After several investigations where team members had experiences and I didn't, my wife suggested that it was because I didn't have an open mind entering the investigation and the spirits knew it. She also said that I would probably never have an experience until I set aside all the "science stuff" and entered the hunt with

Patti Unvericht
Investigator / E.V.P. Specialist

an open mind. Since my wife, having had an uncanny ability to sense spirits since she was a child, seemed to have a good point, I decided to give it a try. I immediately had an experience. During one of our nightly investigations at the Crystal Palace Saloon in Tombstone, Arizona, I actually witnessed a cup fly across the room and crash into a stack of other cups causing an enormous mess which I had to clean up. There was no one else on the saloon floor but me. Everyone else was investigating the basement at the time so I could not explain what had just happened. Even though I still approach investigations with science, I also enter with an open mind. My advice to you is, be skeptical

Dawane Harris
Founder / Lead Investigator

Mind-set

but if you want to have a personal paranormal experience ie: touch, smells, voices and cold spots, enter your investigation with the thought that anything is possible.

Chapter 2
Equipment

The most common misconception formed by someone who wants to begin ghost hunting is that it will be a hobby that is too expensive. Although ghost hunting can be very expensive, I will give you a few tips on how to get started for around $200. First and foremost, no ghost hunter is complete without the coveted voice recorder. I have experimented with several different types and brands of voice recorders and have found that you can purchase a voice recorder online for $19.95 that works just as well for recording E.V.P.'s as a $150 name brand MP3 that you can get in a department store. The difference

Typical Voice Recorder

Equipment

between the two is fairly simple, longevity and accessories. Generally speaking, the more you spend on a voice recorder, the more options it gives you. For example, higher priced MP3's give you options like storing videos and photos. Even though these are options you might use in your personal life, they are not necessary in our field. Since you can download free audio editing systems online, the less expensive voice recorder will work just fine.

Digital cameras are almost essential to any ghost hunter. While doing E.V.P. sessions, someone should be snapping as many pictures as possible. Digital photos can be reviewed after an investigation fairly easily with any free photo editing

Kodak Digital Camera
$108

Equipment

software. By uploading your photos and having the ability to zoom in and really analyze them, the opportunity exists to identify evidence of paranormal activity. You can purchase a digital camera anywhere for around $100.

Next, I would suggest purchasing a K2 meter. This meter is designed to detect electromagnetic field in a specific area that you are investigating. Generally, these meters have different colored L.E.D. to indicate the strength of the field that you have entered. It is thought that spirits generate an electromagnetic field so, with that being said, it seems to be an essential tool for locating possible manifestations. You can generally find K2 meters online for 40 to 60 dollars.

K₂ Meter

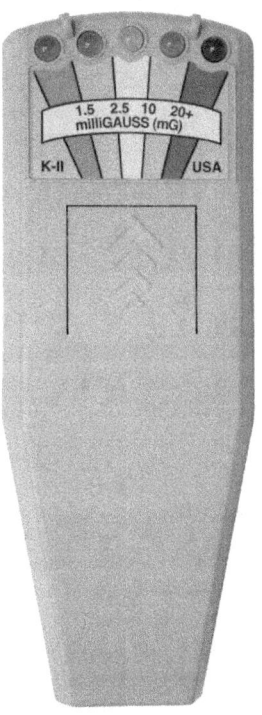

Equipment

As you can see, it is fairly inexpensive to get this hobby started. Most of the things that you need are items most people already have. Obviously, there is a lot more equipment out there that you can use during an investigation but remember, most of the tools used by the teams that you see on the popular television shows are very expensive. The Thermal Imaging Camera or T.I.C., for example, retails for around $7500.00 which is not an amount that most teams are willing or able to pay. Remember, ghost hunting is a hobby. Less than one percent of paranormal groups make a living ghost hunting.

Chapter 3
Picking A Location

There are three main types of locations that ghost hunters will approach for an investigation. The first is a private residence. Although someone's home can be an exciting hunt depending on the type of activity it has had, it can also be a very difficult situation to deal with. Generally speaking, when you are dealing with a private residence, you are dealing with someone's private life which can be very intimate. When someone asks you to do an investigation of their home, most of the time they expect you to keep in confidence all personal conversations and findings. Some people might be embarrassed,

Location
Greece, New York

Picking A Location

confused and scared of what you may find. With that in mind remember to be careful when talking about your experiences and don't talk about anything that may be confidential. On the other hand, when you are hunting a commercial location or business, most of the time the owners want you to tell as many people as possible. It seems that today, no matter what station you turn your television to you will run into some sort of ghost hunter show. Merchants understand the value of exploiting the "in" thing and most of the time they want their location to be haunted. This poses another problem, make sure you never over exaggerate a location's activity just to help a business owner "prove" that

Valentown
Victor, New York

Picking A Location

their place is haunted.

There are literally hundreds of possible haunted locations around the country that you can pay to investigate. Hunting these locations will eliminate any confidentiality issues you may encounter at private homes and feeling obligated to say a business is haunted when you feel otherwise. Most of the time, these locations will grant you permission to post your findings on your websites and other places as part of the price you pay to do the investigation. These locations are a great starting hunt location for beginners.

Picking a location can be one of the most exciting and fun parts of a ghost hunters job. There is an old saying,

Crystal Palace Saloon
Tombstone, Arizona

Picking a Location

"The anticipation is the most exciting part". Nothing could be closer to the truth. Countless times we, as a group, have chosen a location months prior to the investigation and the wait just about drove us insane.

Rolling Hills Asylum
East Bethany, New York

Chapter 4
The Walk-through

Before you start your investigation, it is always recommended that you and your team perform a walk-through with the owner. The walk-through is important because this allows you to pinpoint activity hotspots.

During the walk-through, you should be prepared to mark "hotspots" in some way. I like the black tape method. Get a roll of black electrical tape and whenever you come to a "hotspot" mark it with an X. We do this to identify locations that we are going to do in depth E.V.P. sessions or set up night vision cameras.

During your interview with the owner, be sure to ask as many questions as

The Walk-through

The Walk-through

possible about the history of the location. You may need to do your own research prior to arriving for your investigation. It is important to know as much as possible about the location, this will enable you to use "trigger" objects during your investigation.

Trigger objects are an important tool in any investigation. These objects are generally things that a ghost or spirit would have held very close to themselves. For example, if you are dealing with a small child, try using a doll or toy to instigate activity. Trigger objects have been known to draw out and even help manifest spirits during an investigation. The rest of your walk-through should be spent figuring out

Shock Therapy Room
Rolling Hills Asylum

The Walk-through

how your team will split up to do the investigation. Be sure to split the teams evenly and don't ever investigate by yourself. If you have at least two people per team you can verify activity.

Chapter 5
The Investigation

Here we go! Possibly the most exciting thing you will ever take part of in your life is about to begin. There are a lot of people out there that claim to be "professional ghost hunters", the fact is, there is no such thing. Here is the way I see it, there is no right or wrong way to do an investigation there are only guidelines. As a beginner, you should study several different groups and their procedures to formulate your own technique that fits you and your comfort level. Whatever process you decide to adopt, don't ever allow anyone or any group to tell you that you are doing it wrong. This being said, there are a few

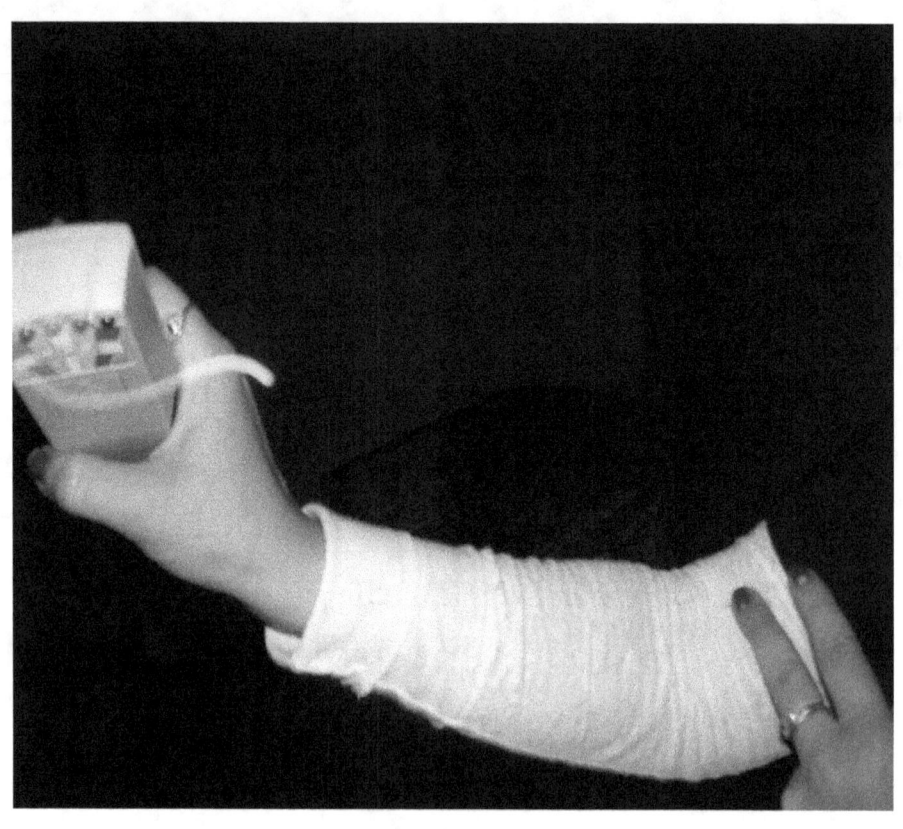

Lead Investigator, Denise Harris, uses a K2 Meter during an investigation.

guidelines that you should always follow. First and foremost, don't ever, for any reason, fake or alter evidence to help a client or yourself gain notoriety. Not only will this eventually be uncovered, but it will most definitely ruin any good reputation that you have made for yourself.

The other guideline that you should follow at all times is the no charge rule. In our field, we, as investigators, have an obligation to help people that call upon us for free. If this is going to be a problem for you, this obviously is not the right field for you. Remember, most of the clients you will come in contact with are emotionally distressed, probably to the point that they are willing to pay you.

Set up to investigate

The Investigation

Since it is not morally correct to take advantage of people in need, we offer our services free of charge to all clients.

Just starting out, you will probably be using a digital voice recorder, a camera and maybe a K2 meter. It is better to do your first hunt using the basics. This allows you to concentrate more on your surroundings than trying to keep track of too much equipment.

Once you have set up your base of operations, a central location where you and your team can meet throughout the night, send your team to their predetermined locations and begin their E.V.P. sessions. E.V.P. stands for Electronic Voice Phenomenon. The human ear has a range of 20hz to

Adams Basin Inn

Founder, Dawane Harris, Explains how E.V.P. sessions are conducted.

The Investigation

20,000hz (hz=hertz), whereas some of the upper end digital voice recorders have a range of 5hz to 30,000hz. Voices from beyond generally can't be heard with the unaided ear, but if you press record on a voice recorder, you will pick up sounds that you would have otherwise not heard. A perfect example of this is a dog whistle. If you blow a dog whistle, most humans cant hear the frequency but if you play back a recording of one, you will almost always hear it.

E.V.P. sessions are fairly simple to do. Always start by noting the time and date followed by the names of the people in the room with you. This will verify the number of people present during the

Lead investigators are responsible for training and explanations.

review process. You should have gotten a history of the location and formed questions pertaining to that history. With a good knowledge of what may have occurred at your location, you should get decent responses on your voice recorders. Always allow 7-10 seconds between questions, this gives enough time for any response you may get. Asking questions isn't as hard as you may think. The way I look at it is, if I were to be face to face with a ghost, what would be the one thing I would want to know about him or her. Once you have that question figured out, follow that same line of questioning for the remainder of the session.

During your E.V.P. sessions, you should

Patti, Denise and Dawane on the Hunt!!!

The Investigation

be monitoring your K2 meter and taking as many pictures as possible. Photos have proven very valuable in providing evidence to clients. There are countless photos that have been published around the world that are claimed ghost photos. The only thing better than capturing an apparition in a photo is capturing one on video.

Throughout your investigation you will most likely come across some pretty intense situations. Remember, no hunt is ever the same as another. You will continually run into new situations and it is important that you learn from your mistakes.

Equipment Tech. Mark Unvericht

Chapter 6
The Review

Now that you have completed your investigation, you can concentrate on the hard part. Reviewing all the evidence you have gathered from an investigation can be a very daunting task especially if you don't have anyone to help you. I would suggest that you appoint at least 2 people in your group to help you with the review process. Having more than one person reviewing material has its benefits. First of all, while listening to E.V.P.'s, it is always a good idea to have an extra set of ears to verify any suspected voice or sound that you uncover. The most important thing that having more than one reviewer does is it reduces the time it

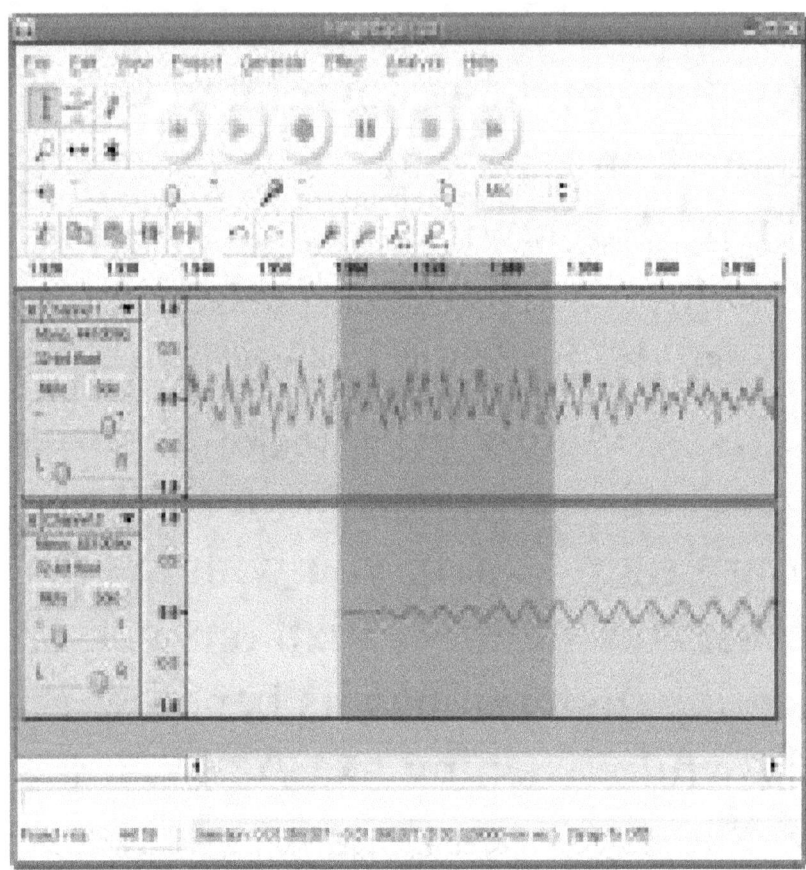

Audacity is a free audio editing software.

The Review

takes to review evidence greatly. The last thing that any investigator wants is to be glued to a set of earphones and a computer for any extended period of time. The more people that volunteer to review, the less time it will take, obviously.

While reviewing E.V.P.'s, be careful to not mistake ambient noise, human breathing, growling stomach and whispering for sounds of interest. This is why it is extremely important for everybody involved with the session to explain out loud any normal noise that may occur during the recording process.

If you happen to discover a sound of interest, I would suggest that you amplify the E.V.P. with one of the several free

Woman in the Mirror

The Review

audio editing programs offered online.
Once you have edited the sound file, you
can then save the clip to be uploaded to
your website or playback device for
sharing with your client.

In our group, we have a member who
loves to review our pictures taken during
the hunts. The way she approaches it is
quite simple. She will upload the photos
to her computer and use a photo editor to
magnify each picture to scan for
anomalies. This allows her to discover
things that otherwise would never be
seen in a regular sized photo. It is a very
meticulous process but is very effective.

When you get to the point of using
video and other recording devices, you

will need to figure review time for that also.

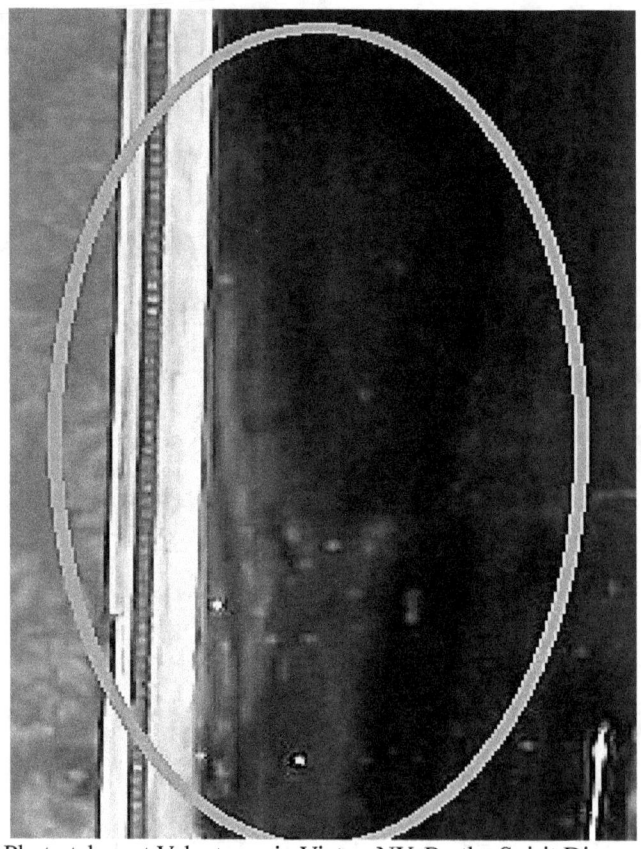

Photo taken at Valentown in Victor, NY. By the Spirit Diggers

The Review

Using night vision D.V.R. systems and other infrared video recording devices add a whole new excitement to your investigation and review.

Photo taken at the Crystal Palace Saloon in Tombstone, Arizona

Chapter 7
The Haunts

Since there are basically four different types of haunts, now would be a good time to explain them to you. The first type of haunt is the demonic haunt. Demons have been on earth since the beginning of time. They only have one thing on their mind, possessing a human being. Demons have been called Legion, fallen angels, black angels and angels of death. Whatever you call them, they are nasty and you definitely don't want to mess with them. Classic tell tale signs of a demonic haunt are rancid smells and extreme increase in room temperature. If you suspect a location to have a demonic haunt, do not try to confront it. Advising your client to contact the Catholic church

Demon

is the best thing to do.

The second type of haunt is a poltergeist, which in German means noisy ghost or spirit. These are the types of ghosts that will tip over appliances, throw things around the house and hide your keys. Even though a poltergeist is not as bad as a demon, they can be just as scary. Unlike a demon who haunts a person, a poltergeist haunts a location and can be left behind by moving to a different location.

Number three on the list is an intelligent haunt. These spirits will cast shadows, be vocal and turn your lights on and off. Since they are simply trying to make contact, there is generally no reason to be

An illustration that shows Poltergeist activity

scared of them. Most of these harmless spirits are simply trying to say goodbye to a loved one or relay some sort of message. On some occasions, the ghost doesn't know he or she is dead and is just continuing their life as if nothing is different.

The final type of haunt is the most common. The residual haunt is like an imprint in time. Like a broken record, these ghosts keep repeating the same thing over and over again. Generally replaying the last few moments of their lives, the spirit has absolutely no awareness of their surroundings. A perfect example of a residual haunt is the Gettysburg Battlefield.

Gettysburg Battlefield

Chapter 8
Revealing Evidence

The most exciting and anxious time for your client will be the revealing process. This is where you sit down with the client and go over all the evidence that you and your team collected during the investigation. It is important that you show only the evidence that you have scrutinized several times and that you can say for a fact that you have no logical explanation for. Not only will this put the client's mind at ease but it will make you feel better knowing that there is no possibility that you inadvertently gave false evidence.

The first thing I reveal to my client are the E.V.P.'s that we capture. It is always

Explain how double exposures can sometimes indicate paranormal activity

Revealing Evidence

fun to see the expression on the face of someone who has just heard a voice believed to come from beyond the grave. Next comes the photo evidence. Although it is very rare to capture an apparition in a photo, it does happen. Showing one of these photos to a client is very rewarding for all involved. Once you are using video, this would be a good time to show any clips of interest to your client.

   The final part of revealing evidence is classifying the type of haunt you believe your client to have. Be sure to fully explain all the different types of haunts in detail and which one you believe them to have and why. You may also want to give advice on the way to deal with whatever

Ask for something to move, you never know what will happen!!!

Revealing Evidence

type of haunt they have by sharing different website addresses and authors that specialize in their type of haunt. Always make yourself available to a client after an investigation for any further questions or concerns.

Chapter 9
Personal Experiences

Over the past few , I have had numerous personal experiences with what some would call ghosts. I am still a skeptic even though I have been in some pretty scary and unexplainable situations.

On one occasion, while investigating the Crystal Palace Saloon in Tombstone, Arizona, I experienced what sounded like a choir singing on our enhanced audio system. Wanting to know if I was hearing what I thought I was, I called over a couple of team members to listen also. I did not tell them prior to listening to the audio what I was hearing, but they both agreed it sounded like church music. This audio continued for about 20 minutes and

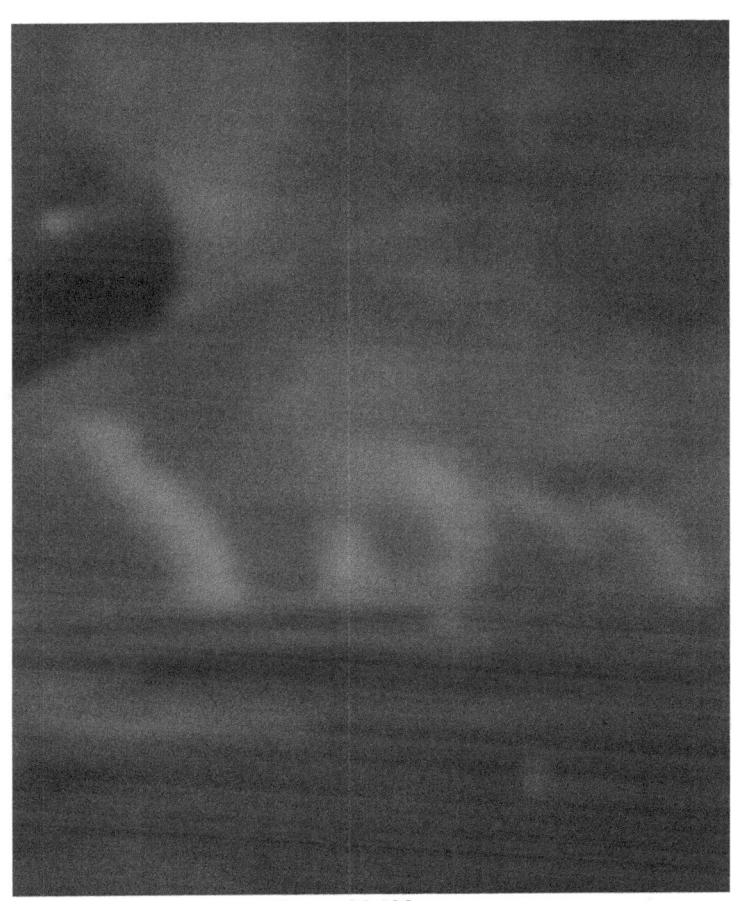

Ghost Girl??

then abruptly stopped. We were sure to start recording the sounds as soon as we all agreed on what we thought it was. The really strange thing was that when we went to play it back for someone, all we could hear was loud static. None of us had any explanation for why this occurred but we all agreed that we definitely heard music.

On a different investigation at the same location a few months later, I was again listening to enhanced audio when a whole shelf of cups fell to the floor by themselves. Just the cups, not the shelf. After gaining my composure and changing my shorts, I inspected the shelf to see why the cups would have fallen. The shelf was about 5 feet off the floor

Crystal Palace Saloon
Tombstone, Arizona

and the cups were stacked 5 deep and 10 across. I could have accepted the possibility that they were stacked crooked and one or two tipped over by themselves and fell off but not every single cup at once. I have absolutely no explanation for 50 cups falling off a shelf all at once.

During a home investigation in Idaho, I was returning upstairs from the basement when it felt as if I was grabbed by the ankle. I tried to tug away from it, knowing it was one of my team members behind me messing around. I ended up losing my balance and falling part way down the stairs. Once I got to my feet, I turn to give someone some choice words. To my surprise, there was nobody there.

Sometimes even children sense a presence!!

The closest person to me was in the basement, down a hallway and in a finished bedroom. Needless to say, I was slightly freaked out. I think this was the beginning of the belief stage for me.

Numerous times I have witnessed lights go on and off by themselves, heard doors slam, footsteps, laughter, music and even crying. After all this, I still think my creepiest experience was at the William Phelps General Store and Palmyra Historical Museum in Palmyra, New York. During our investigation, Mark and I were joined by another gentleman that was affiliated with the location we were at. All three of us were climbing the stairs to the 2nd floor of the Museum.

Personal Experiences

We all, at that point hadn't experienced much activity but that was about to change. As we reached the landing on the second floor, we were suddenly hit with a blast of air and a very loud, what seemed to be a growl coming from the Civil War room. The situation scared me so bad that I jumped backward almost to the point that Mark, who was behind me, thought I was going to fall down the stairs. After discussing what had just happened, we moved into the Civil War room for an investigation.

About the Author
Dawane Harris

Dawane Harris was born February 19th, 1973 in Riverton, Wyoming. After graduating high school in Oregon, he attended the University of Iowa where he obtained a B.S. In Biology.

Having a passion for powerlifting, wrestling and paranormal investigating, his devotion to his family comes first.

Dawane now resides in Spencerport, New York and continues his paranormal research.

www.ingramcontent.com/pod-product-compliance
Lightning Source LLC
Chambersburg PA
CBHW070836310526
45788CB00017B/1463